MATHEMATICAL TRICKS THAT WILL MAKE YOU MASTER IN MATHS

Muhammed Sabiq

© 2032 Sabiq V

All rights reserved. No part of this publication may be reproduced, distributed, or transmitted in any form or by any means, including photocopying, recording, or other electronic or mechanical methods, without the prior written permission of the publisher, except in the case of brief quotations embodied in critical reviews and certain other noncommercial uses permitted by copyright law.

For more information, contact the author at:
Email: sabiqvkd@gmail.com

DEDICATION

This book is for every dreamer, learner, and problem-solver who refuses to be held back by challenges.

To the students who reach for five stars, to the professionals who work numbers like a maestro, to the inquisitive minds who solve the mysteries of mathematics you motivate us to make the world invasive and resolve to make the world simpler.

To the educators and mentors who inspire a passion for learning, and to the friends and family who cheer us on at every stage in our development, your support nurtures growth and progress.

Ultimately, this book is for believers, the people who believe that with the right attitude and a couple of clever tricks, everything, even math, can be an adventure.

WHY YOU SHOULD READ THIS BOOK

As a universal skill, mathematics is almost a requirement for every life aspect, whether it is solving problems daily or seeking academic achievements or career goals. But for many people, math is a formidable challenge. Mathematical Tricks That Will Make You a Master in Math got your back, with some easy but effective techniques that will help you face number like a pro.

This book isn't only about learning math — it's about mastering it. Whether you are a student looking to ace your exams, a corporate employee working with numbers, or anyone looking to sharpen your mental flexibility, the tricks and techniques mentioned here will enable you to arrive at answers faster, think laterally, and tackle problems with clarity.

This Book Breaks down even the hardest topic with evidences and step by step explanation. You'll discover how to do speed calculations, solve problems accurately and understand how the tricks work, making them easy to remember and apply.

When you're done with the course, not only will your mathematical skills be amplified, but your entire outlook on problem-solving will be altered. If the sight of math intimidates you, this book will help you unlock its magic and discover how fun and powerful working with numbers can be.

TABLE OF CONTENTS

CONTENTS	Page No.
INTRODUCTION	6
Chapter 1: DIVISION TRICKS	8
Chapter 2: MULTIPLICATION TRICKS	18
Chapter 3: SQUARING TRICKS	25
Chapter 4: MISCELLANEOUS TRICKS	31
CONCLUSION	38

INTRODUCTION

Mathematics can feel daunting at times, but what if there was a way to make it fun, engaging, and remarkably easy? Welcome to Mathematical Tricks That Will Make You a Master in Math, a book designed to transform the way you approach numbers. Whether you're a student aiming for higher grades, a professional dealing with numbers daily, or simply a math enthusiast, this book has something to offer for everyone.

Mathematics is not just about solving equations—it's about thinking creatively and finding patterns that simplify the process. This book focuses on those hidden shortcuts and clever methods that make calculations not only faster but also more enjoyable.

Here's what you'll find in the pages ahead:

- **Division Tricks**: Struggling with long division? These techniques will show you how to break down complex problems into simpler steps, helping you solve them with confidence.
- **Multiplication Tricks**: Learn methods to multiply large numbers swiftly, making even the most intimidating problems manageable. From two-digit multiplications to larger ones, you'll be surprised by how simple it can be!
- **Squaring Tricks**: Squaring numbers can be tricky, but not anymore! These tricks will enable you to calculate squares of numbers quickly and accurately, perfect for competitive exams or mental math challenges.
- **Miscellaneous Tricks**: Explore an array of additional tips and methods that cover various aspects of mathematics, from finding remainders

to approximations, all tailored to save you time and effort.

This book isn't just about learning math—it's about mastering it. By the end of your journey through these pages, you'll not only solve problems more quickly but also develop a deeper understanding of mathematical concepts.

Get ready to amaze yourself and others with your newfound skills. Let's embark on this exciting journey and uncover the magic hidden in the world of numbers!

CHAPTER 1
DIVISION TRICKS

Although division is one of the more difficult operations on math, there are some maneuvers designed to speed things up and make your life easier. These "tricks" are all about rendering complex problems into simpler, more manageable, chunks and using patterns to eliminate as many steps as possible. The good news is that with practice [easy division becomes some process that can be done without looking in adverse situations.

Division tricks are based on the relationships between numbers, and they are the key to mastering division. These techniques are for avoiding the repetitive long-division methods which can speed you toward the correct answer without as much work. From those dividing very large numbers or decimals to even others finding remainders, these techniques are reliable shortcuts that provide result with accuracy.

In exams, professional tasks, or even daily problem-solving, you might get a situation where you need a calculation really fast, and these division tricks help you tremendously. The reason for doing so is to reduce the amount of work to do a division, so that one could spend time on the big picture while deciding quickly what to do following a data-driven analysis instead of spending excessive time on tens of thousands of data points and their mathematical manipulations.

Otherwise, these tricks are great practice for students — they're good for your division skills and help develop a better understanding of the numerical relationships involved. Mastering this system — and ultimately many more number-based methods — will give you confidence and a deeper appreciation for the simplicity and order of division.

TRICK 1: DIVIDING BY 1

A crucial thing to keep in mind is that any number divided by one is the number itself! As an example, 200 divided by 1 equal 200

TRICK 2: DIVIDING BY 2

All even numbers—those that finish in 0, 2, 4, 6, or 8—are divisible by 2.
For example, 444 can be split by 2 = 222

TRICK 3: DIVIDING BY 3

A little-known but interesting fact is that if a number can be divided by three, so can the sum the total of all the numbers combined together!

Here's an illustration: 27 is the number;
2+7 = 9.
9 / 3 equals 3.

Doubting?
Let's examine a larger figure: 333.
3+ 3 + 3 = 9
9/3 equals 3.

Alternative explanation,

- When a number is divisible by three, it can be found using this trick.
- Example: 765

The first thing we do is add up the numbers 7 + 6 + 5 = 18.
Next, we calculate the total of the digits: 1 + 8 = 9

Another Example:
Think about 1965.
The first thing we do is add up the numbers 1 + 9 + 6 + 5 = 21.
Next, we calculate the total of the digits: 1 + 2 = 3.

The original integer is also divisible by three if this number is. finds out if an integer is divisible by three.

TRICK 4: DIVIDING BY 4

Examine the number's final two digits when dividing by 4. Can you divide those two numbers by four? That implies the whole number is, too!

Let's examine a very long number like 345620,
Consider 20 which is divisible by 4
86405 is obtained by dividing 345620 by 4.

Another example is,

1234516
Consider 16 which is divisible by 4
308629 is obtained by dividing 1234516 by 4.

TRICK 5: DIVIDING BY 5

The fact that any number that ends in a 0 or a 5 is always divisible by 5 is another excellent reminder.

Like, 20 is divisible by 5

1234560 is divisible by 5 (the quotient is 246912)

TRICK 6: DIVIDING BY 5

- In order to divide a large number by 5.
- Multiply that number with 2 and move the decimal point.
- For example
  200/5
  200 x2 = 400
  40.0
  Answer is 40.0

  Another example is,
  600/5
  600x2 = 1200
  120.0
  Answer is 120

TRICK 7: DIVIDING BY 6

If a number's digits add up to something divisible by 3 and end in an even number, it will be divisible by 2,3, and 6.

330 divided by 3 is 110

270 divided by 3 is 90

TRICK 8: DIVIDING BY 7

Start with the number's final digit to determine whether it is divisible by 7. Subtract the doubled amount from the number's remaining two digits after you've doubled that amount.

Assume that 175 is the number.
First, get ten by doubling the number five.
7 is the result of subtracting 10 from 17, and it is divisible by 7.

Therefore, 175 is divisible by 7, we might claim. And sure enough, 175 divided by 7 equals 25 when we check out the job!
Another Example is 245
5+5 = 10
24-10= 14 divisible by 7
245 / 7 = 35

TRICK 9: DIVIDING BY 8

Another division trick that works with bigger numbers is this one. The full number will likewise be divisible by 8 if the final three digits produce a number that is.

For instance, the final three digits of 1960, as a separate integer, can be divided by 8,

just like 2840, Divide 2840 by 8 to get 355.

TRICK 10: DIVIDING BY 9

The number itself is divisible by nine if the sum of its digits is also divisible by nine.

For example, 81 is equal to 9 when added together, and when it's divided by nine, the answer is 9.

Another example is,

900
9+0+0= 9 which is divisible by nine
900/9 = 100

TRICK 11: DIVISION TRICK FOR NUMBER 9

- When a number is divisible by nine, it can be found using this trick.
- Example:
Think about 7200.
The first thing we do is add up the numbers 7 + 2 + 0 + 0 = 9.
Next, we calculate the total of the digits: 0 + 9 = 9.

    Another Example:
    Think about 2295.
    The first thing we do is add up the numbers 2 + 2 + 9 + 5 = 18.
    Next, we calculate the total of the digits: 1 + 8 = 9.

The original integer is also divisible by nine if this number is finds out if an integer is divisible by nine

TRICK 12: DIVIDING BY 9

The number itself is divisible by nine if the sum of its digits is also divisible by nine.

For example, 81 is equal to 9 when added together, and when it's divided by nine, the answer is 9.

Another example is,

900
9+0+0= 9 which is divisible by nine

900/9 = 100

TRICK 13: DIVIDING BY 10

A number that ends in a zero is divisible by ten, much like the trick with dividing by five.

Additionally, the number currently in the hundreds, tens, and one's places will have the answer dividers are searching for.

For example, 35 is obtained by dividing 350 by 10.
35 is obtained by dividing 3500 by 100.

TRICK 14: DIVISION TRICK FOR NUMBER 11

- Consider dividing 15,356 by 11.
  The trick to divide by 11, begin by placing a "+" above the farthest right digit.   Therefore, we write "+" above the number 6 in our instance Therefore, in our case, we write "−" above the number 5. In order to do this, we would put a "+". After that, write a "−" above the second-rightmost digit you want to split by 11. over the numbers three and one and a "−" above the number five.

And lastly,
+ - + - +
1 5 3 5 6

We now combine our "+" and "−" so that we can add them together and subtract the two.

As illustrated below, (1 + 3 + 6 -5-5)
10-10 = 0

Thus, our outcome is zero. This proves that the number we suggested is divisible by 11.

TRICK 15: SUPER SIMPLE DIVISIBILITY RULES

You've got 210 pieces of cakes and want to know whether or not you can split them evenly within your group. Rather than whip out the calculator, use these simple shortcuts to do the math in your head:

- ✓ Divisible by 2 if the last digit is a multiple of 2 (210).
- ✓ Divisible by 3 if the sum of the digits is divisible by 3 (522 because the digits add up to 9, which is divisible by 3).
- ✓ Divisible by 4 if the last two digits are divisible by 4 (2540 because 40 is divisible by 4).
- ✓ Divisible by 5 if the last digit is 0 or 5 (9905).
- ✓ Divisible by 6 if it passes the rules for both 2 and 3 (408).
- ✓ Divisible by 9 if the sum of the digits is divisible by 9 (6390 since 6 + 3 + 9 + 0 = 18, which is divisible by 9).
- ✓ Divisible by 10 if the number ends in a 0 (8910).
- ✓ Divisible by 12 if the rules for divisibility by 3 and 4 apply.

Example: The 210 slices of Cakes may be evenly distributed into groups of 2, 3, 5, 6, 10.

TRICK 16: RATIO CROSS-MULTIPLICATION

- Cross-multiplying fractions makes division easier.

    Examples: 45/9 =5

    27/3 =9

TRICK 17: RATIO CROSS-MULTIPLICATION

- Cross-multiplying fractions makes division easier.

Examples: 45/9 = 5

27/3 = 9

TRICK 18: DIVIDING USING RECIPROCALS

For simpler computations, convert division to multiplication by the reciprocal.

Example,
$35 \div 5 = 35 \times 1/5 = 35 \times 0.2 = 7$

TRICK 19: DIVIDING OF EVEN NUMBERS

To make calculations easier, divide by an even number by half of the divisor and the dividend.

Example,
$528 \div 22$

Half the numbers and divide,
$264 \div 11 = 24$

TRICK 20: GETTING RESULT BY MULTIPLICATION

To expedite the operation, compute multiples of the divisor beforehand.
Example,

$55 \div 11$

Multiples of 11 will be 11, 22, 33, 44, 55, 66

$55 = 5 \times 11$

Answer is = 5

TRICK 21: EASY LONG DIVISION

Pay attention to the dividend and divisor's initial few numbers. Step-by-step, repeat

Example,

$264 \div 8$
Start with left digit $26 \div 8 = 3$ remining is (26-24=2)
Drop the following number.,
$24 \div 8 = 3$

Answer is = 33

TRICK 22: FRACTION SIMPLIFYING

Eliminate common components from the dividend and divisor to simplify them.

Example,

$88 \div 66$

Simplifying = 88/66 = 8/6 = 4/3 = 1.3

CHAPTER 2
MULTIPLICATION TRICKS

Multiplying numbers can be quite a tedious task, but with a few clever tricks and methods, it can become much easier. The methods aim to simplify the multiplication of big numbers by dividing them into smaller, manageable components or discerning patterns that accelerate the calculation. After a bit of practice, these strategies become second nature, making what feel like impossible problems effortless.

These tricks are mainly aimed at helping you see multiplication more intuitively. Now look, you can do the calculations very quickly and no need of calculator, if you find the trick and logic behind the summation. By minimizing the use of pen and paper, these techniques help save time and improve mental math skills, so you can be more confident to solve the problems using few to no tools.

Also, these times up tricks have a high utility and can be used in different type of situations right from competitive exams to day to day situations where you can do some quick calculations. This makes them especially helpful for people in professions where they use lots of numbers, who want to save time and make sure they are correct.

When you master these tricks, you're not just learning faster ways to calculate — you're developing a new way of seeing math. Through recognizing patterns understanding the underlying logic of multiplication, you'll unlock the potential to solve problems creatively and efficiently, turning what was once a tedious task into an enjoyable and rewarding skill.

TRICK 1: MULTIPLY BY 4

In order to multiply any number with 4.
Multiply that number with two, then multiply by two again, then add it.
For example,
300x4
(300x2) + (300x2) = 600 + 600 = 1200
Answer is 1200

Another Example,
800x4
(800x2) + (800x2) = 1600 + 1600 = 3200
Answer is 3200

TRICK 2: MULTIPLICATION TRICK

In order to multiply two numbers and one of them is even number.
Easily subdivide to get the answer
For example
600 x 15
300 x 30..........600/2 & 30x2
6 x 1500..........300/50 & 30x50
Answer is 9000

Another example is,
300 x 15
150 x 30
3 x 1500
Answer is 4500

TRICK 3: CRISS CROSS MULTIPLICATION

In order to square any two-digit number we will use this method

For example, if you want to square 54^2

Multiply 5x5 = 25

Multiply 4x4 = 16

Multiply 5x4x2 = 40

Then write down the results as below,

2516 +

  40

2916 will be the answer.

Note: If the square is single digit number lets write zero before the number, for example for the $2^2 = 4$, we will write 04

TRICK 4: MULTIPLY ANY TWO DIFFERENT TWO-DIGIT NUMBERS

In order to multiply any two-digit numbers we will use this method

For example, if you want to multiply 42 x 23

4 2

2 3

Multiply right-side digits, 2 x 3 = 6

Cross multiply and add = (4x3) + (2x2) = 16

Multiply left side digits, 4 x 2 = 8

Then write down as below,

966 (from 16 we will write only 6 and the balance 1 we will add with 8)

TRICK 5: MULTIPLY BY 5

In order to multiply any number with 5.
Take half of that number, if answer is whole, add a 0 at the end. if it is not, add a 5 at the end.
For example,
530 x 5
    265
    2650
Answer is 2650

Another example,
111 x 5
55.5
Answer is 555

TRICK 6: MULTIPLYING BY 6

If you multiply 6 by an even number, the answer will end with the same digit. The number in the ten's place will be half of the number in the one's place.
Example: 6 x 2 = 12.

TRICK 7: MUTILPY 9 WITH (1 TO 10)

In order to multiply 9 with digits from 1-10
Write down from 0-9 on the left side from top to bottom & write 0-9 on the right side from bottom to top
1 x 9 = 09
2 x 9 = 18
3 x 9 = 27
4 x 9 = 36
5 x 9 = 45

6 x 9 = 54
7 x 9 = 63
8 x 9 = 72
9 x 9 = 81
10 x 9= 90

TRICK 8: MUTILPYING TWO NUMBERS NEAR TO 100

In order to two numbers near to 100.
Follow below Steps-
Example,
93 x 94
(100-93) & (100-94) ……. Deduct both numbers from 100
7 & 6…………Write down those numbers
Add 7+6 = 13 ……… (100-13) =87…...Deduct the sum from 100
Multiply 7 x 6= 42
Answer is 8742

Another Example is,
97x98
(100-97) & (100-98) ……. Deduct both numbers from 100
3 & 2…………Write down those numbers
Add 3+2 = 5……... (100-5) =95…....Deduct the sum from 100
Multiply 3x2 = 06…...because of single digit write zero on left side
Answer is 9506

TRICK 9: MULTIPLY BY 99

In order to multiply any number with 99.
To obtain the desired outcome, first multiply the number by 100, and subsequently subtract the original number from this result.
Example,
Multiply 11 x 99
(11 x 100) – 11
1100 – 11
Answer is 1089

Another Example,
21 x 99
(21 x 100) – 21
2100 – 21
Answer is 2079

TRICK 10: MULTIPLY LARGE NUMBERS

To easily multiply two double-digit numbers, use their distance from 100 to simplify the math:
Subtract each number from 100.
Add these values together.
100 minus this number is the first part of the answer.
Multiply the digits from Step 1 to get the second part of the answer.

TRICK 11: FINGER MULTIPLICATION TABLES

Everyone knows you can count on your fingers. Did you realize you can use them for multiplication???

A simple way to do the "9" multiplication table is to place both hands in front of you with fingers and thumbs extended. To multiply 9 by a number, fold down that number finger, counting from the left.

Examples: To multiply 9 by 5, fold down the fifth finger from the left. Count fingers on either side of the "fold" to get the answer. In this case, the answer is 45.

To multiply 9 times 6, fold down the sixth finger, giving an answer of 54.

CHAPTER 3
SQUARING TRICKS

Squaring numbers is usually a long road — especially when they are big. But there are techniques that make squaring numbers a lot faster and easier. This book covers tricks that you can use to make squaring easier — including tricks that rely on pattern recognition and using shortcuts that minimize the steps you take to calculate the right answer.

These squaring tricks aren't just useful for higher-speed computation — they help you make sense of the underlying structure of numbers, which is what allows you to handle even complicated squaring problems. You can use these techniques to square numbers in your head without actually squaring it the traditional way.

These techniques can be especially helpful in competitive examinations, mental maths competitions, or whenever there is less time and accuracy is important. They enable you to answer problems simply, raise up your confidence, and sharpen your speediness under pressure.

Mastering the tricks for squaring numbers enhances both your mathematical agility and your problem-solving skills. With consistent practice, squaring numbers will no longer be a daunting task but an opportunity to showcase your newfound expertise in the art of math.

TRICK 1: SQUARING

- To implement this straightforward technique, it is necessary to adjust the equation in such a way that the units digit becomes zero. This modification simplifies the multiplication process, as calculations are more manageable when the units digit is zero.
- Example Finding Square of 55

  (55+5) (55-5) + 5x5

  60 x 50 + 25

  600 x 5 + 25

  3000 + 25

  Answer is 3025

  Another Example,

  11^2

  (11+9) (11-9) + 9 x 9

  20 x 2 + 81

  40 + 81

  Answer is 121

TRICK 2: TAKING SQUARE OF ANY TWO-DIGIT NUMBER END IN 5

In order to find square of any two-digit number end in 5 we will use this method

For example, 15

- ✓ Take square of right-side digit, 5 = 25
- ✓ Multiply the left side digit with its next number = 1 x 2 = 2
- ✓ Then write down as below,

Answer will be = 225

Alternative Explanation is,

Use this equation $(n5)^2 = n \times (n+1)$ and add 25 at the end.

Considering same example, 15

n = 1
1 (1+1) = 1 x 2 = 2
Answer will be 225

TRICK 3: TAKING THE SQUARE OF A VALUE NEAR 50

Let's look at the numbers that fall between 41 and 59, which are near 50. The following formula is used to determine the square of these numbers:

1. Deduct 25 from the total. The first two digits of the final outcome will be this number.
2. Determine the difference between the number and 50.
3. The final result's final two digits will be the result of squaring that difference.

Example, 48^2
- ✓ Step 1 = 48-25 = 23
- ✓ Step 2 = 50-48 = 2
- ✓ Step 3 = 2^2 = 4

The number's last two digits are 04, while its first two are 23.

$$48^2 = 2304$$

Another Example is, 57^2
- ✓ Step 1 = 57-25 = 32
- ✓ Step 2 = 57-50 = 7
- ✓ Step 3 = 7^2 = 49

The number's last two digits are 49, while its first two are 32.

$$57^2 = 3249$$

TRICK 4: SQUARE USING SQUARE FORMULA

Formula is, $(a+b)^2 = a^2 + 2ab + b^2$

Example, 49^2

- ✓ Split 49 into 40 + 9, a = 40 & b = 9
- ✓ Add in equation
  $40^2 + (2 \times 40 \times 9) + 9^2$
  $1600 + (720) + 81$
  2401

Another Example is, 45^2

- ✓ Split 45 into 40 + 5, a = 40 & b = 5
- ✓ Add in equation
  $40^2 + (2 \times 40 \times 5) + 5^2$
  $1600 + (400) + 25$
  2025

TRICK 4: SQUARING BY SIMPLE ADJUSTMENT

Adjust and compute for numbers that are close to a round number.

Formula is, $(a-b)^2 = a^2 - 2ab + b^2$

Example, 98^2

- ✓ 98 = 100 - 2, a = 100 & b = 2
- ✓ Add in equation
  $100^2 - (2 \times 100 \times 2) + 2^2$
  10000 - (400) + 4
  9604

Another Example is, 96^2

- ✓ 96 = 100 - 4, a = 100 & b = 4
- ✓ Add in equation
  $100^2 - (2 \times 100 \times 4) + 4^2$
  10000 - (800) + 16
  9216

TRICK 5: GENERAL TRICK FOR SQUARING OF TWO DIGIT NUMBERS

Write number into 10a and unit (b), then use below equation,

$(10a+b)^2 = 100a^2 + 20ab + b^2$

Example, 98^2

- ✓ a = 9 & b = 8
- ✓ Add in equation
- ✓ $100a^2+20ab+b^2$
  $100 \times 9^2 + (20 \times 9 \times 8) + 8^2$
  $8100 + 1440 + 64$
  9604

Another Example is, 96^2

- ✓ a = 9 & b = 6
- ✓ Add in equation
- ✓ $100a^2+20ab+b^2$
  $100 \times 9^2 + (20 \times 9 \times 6) + 6^2$
  $8100 + 1080 + 36$
  9216

CHAPTER 4
MISCELLANEOUS TRICKS

This book uses gravity-defying tricks to amaze and amuse the reader, and to demonstrate the importance of mathematical patterns and logic. These tricks transform simple mathematical operations into humorous and mind-boggling ends. They demonstrate how numbers, with a little creative thinking, can expose connections that allow calculations to be done more quickly, easily and surprisingly.

At the heart of these magic tricks are predictable sequences of numbers, which you can use to confidently predict outcomes that seem impossible. For instance, you can perform any number followed by a series of operations, and you always arrive at the same answer at the end — often it's a fixed numerical value, or a series, or an answer that appears nonsensical. These tricks work due to intrinsic properties of the numbers and operations applied on it which means they are always true.

The beauty of these miscellaneous hacks is their simplicity. It has been learned about simple methods and it simply needs some steps to be followed by people who have some basic knowledge about mathematics. It means they are accessible to all ages and all levels of mathematical ability. And they are not just useful for impressing others; they can also help your mental math and deepen your understanding of mathematical reasoning.

In general, these tricks illustrate how numbers can have surprising properties, and that we can attain surprising outcomes via manipulating them in elementary ways! Used to amuse, instruct, or amplify your problem-solving prowess, these tips and tricks prove that math can be entertaining and practical. So, try these out, and before long you will be viewing numbers in a fresh new way — full of patterns, surprises, and infinite possibilities.

TRICK 1: THINK OF A NUMBER

- Pick a whole number between 1 and 10.
- Add 2.
- Multiply by 2.
- Subtract 2.
- Divide by 2.
- Subtract your original number.
- Everyone's final answer will be 1

TRICK 2: SAME THREE-DIGIT NUMBER

- Think of any three-digit number in which each of the digits is the same. Examples include 333, 666, 777, and 999.
- Add up the digits.
- Divide the three-digit number by the answer in Step 2.
- Everyone's final answer will be 37

TRICK 3: SIX DIGITS BECOME THREE

- Take any three-digit number and write it twice to make a six-digit number. Examples include 371371 or 552552.
- Divide the number by 7.

- Divide it by 11.
- Divide it by 13.
- The answer is the same three-digit number.

TRICK 4: THINK OF A NUMBER

- Choose a number from 1 to 8.
- Multiply it by 2.
- Now multiply by 5.
- Subtract 5.
- Finally add 7.
- The first digit is the number you chose and the second digit is the number 2.

TRICK 5: THINK OF A NUMBER

- Think of any number.
- Double the number.
- Add 9 with result.
- Subtract 3 with the result.
- Divide the result by 2.
- Subtract the number with the first number started with.
- The answer will always be 3.

TRICK 6: THINK OF A NUMBER

- Ask someone to pick a number.
- Add the next higher number to it.
- Add 9 and divide by 2, and then subtract the original number.
- The answer will be 5.

TRICK 7: THINK OF A NUMBER

- Write down a number.
- Add 5.
- Multiply by 3.
- Minus 15.
- Divide by your original number.
- Add 7.
- Your answer will always be 10

TRICK 8: THINK OF A NUMBER BETWEEN 20 AND 100

- Pick a number between 20 and 100.
- Now add your digits together.
- Now subtract the total from your original number.
- Finally, add the digits of the new number together.
- Your answer will always be 9

TRICK 9: THINK OF A 3-DIGIT NUMBER

- Pick a 3-digit number with 3 different digits.
- Now reverse the digits to get a new number.
- You now have two numbers. Subtract the smaller number from the larger number.
- Now add up the digits of the result.
- The answer will always be 18

TRICK 10: THINK OF A 3-DIGIT NUMBER

- Pick a 3-digit number with all three digits different.
- Reverse the digits and subtract to get another 3-digit number.
- Reverse the digits of the difference and add it to the difference.
- Your sum will be 1089.

For example, start with 845. Then 845 − 548 = 297 and 297 + 792 = 1089.

TRICK 11: THE 11 RULES

This is a quick way to multiply two-digit numbers by 11 in your head.

- Separate the two digits in your mind.
- Add the two digits together.
- Place the number from Step 2 between the two digits. If the number from Step 2 is greater than 9, put the one's digit in the space and carry the ten's digit.

Examples: 72 x 11 = 792.

57 x 11 = 5 _ 7, but 5 + 7 = 12, so put 2 in the space and add the 1 to the 5 to get 627.

TRICK 12: CONTAINS THE DIGITS 1, 2, 4, 5, 7, 8

- ✓ Select a number from 1 to 6.
- ✓ Multiply the number by 9.
- ✓ Multiply it by 111.

- ✓ Multiply it by 1001.
- ✓ Divide the answer by 7.

The number will contain the digits 1, 2, 4, 5, 7, and 8.

Example: The number 5 yields the answer 714,285.

TRICK 13: MEMORIZING PI

To remember the first seven digits of pi, count the number of letters in each word of the sentence:

"How I wish I could calculate pi."

Pi = 3.141592

TRICK 14: 6 DIGITS BECOME 3

- ✓ Take any three-digit number and write it twice to make a six-digit number. Examples include 123123 or 456456.
- ✓ Divide the number by 7.
- ✓ Divide it by 11.
- ✓ Divide it by 13.

The order in which you do the division is unimportant

The answer is the three-digit number.

Examples: 123123 gives you 123 or 456456 gives you 456.

1. A related trick is to take any three-digit number.
2. Multiply it by 7, 11, and 13.

The result will be a six-digit number that repeats the three-digit number.

Example: 123 becomes 123123.

This works because you're effectively dividing by 1,001.

TRICK 15: SAME THREE-DIGIT NUMBER

1. Think of any three-digit number in which each of the digits is the same. Examples include 111, 222, 333, and 444.
2. Add up the digits.
3. Divide the three-digit number by the answer in Step 2.

The answer always will be 37.

This works because the sum of the digits will always be 3 times the digit, and each three-digit number is a multiple of that digit and 111. Thus, in the last step, you're effectively cancelling out the digit in the division and left with 111 divided by 3, which is 37.

TRICK 16: THE ANSWER IS 2

1. Think of a number.
2. Multiply it by 3.
3. Add 6.
4. Divide this number by 3.
5. Subtract the number from Step 1 from the answer in Step 4.

The answer always will be 2.

This works because you're cancelling out the number you thought of when you subtract it in the last step. You also cancel out the second step when you divide by three, so the only thing you're left with is 6 divided by 3, which is 2.

CONCLUSIONS

With the right techniques, math does not have to be hard. In this book **"Mathematical Tricks That Will Make You a Master in Math"** you have learned some unconventional ways that make life easier and fun in solving mathematical problems. And by refining these techniques, you have sharpened your skill with handling numbers, as well as finding the beauty of patterns and logic within mathematics.

The tricks and shortcuts you've been taught—whether it's to division, multiplication, squaring or other topics—are not just ways to divide the time it takes to get to an answer. These will help you build confidence in how good you are at maths, and help you appreciate the subject a lot more. Whether you're studying for exams, working through everyday chores or investigating math as a hobby, these techniques put you in a position to attack problems with clarity and creativity.

But this book is only a starting point. Consistency and application lead to true mastery. Try them out in your own life, tell someone else about them, and keep looking for new ways to simplify math challenges. By putting what you've learned into action, not only will you sharpen your skills, but you'll encourage everyone around you to view math in an entirely new and imaginative way.

And keep in mind, the mathematics applied here is not just about solving problems — it is a way to develop critical, logical, and creative thinking. And with these techniques and your curiosity, you're ready to tackle any number you find with confidence. Numbers are everywhere and learning what each and every one means will help propel you in your career, academically, or within your personal life.